COMMERCIAL
ORANGE FLESHED
SWEET POTATO
FARMING

John Abraham Godson
& Esonu Nduka Udeala

Commercial Orange Fleshed Sweet Potato Farming

A Concise Practical Guide to OFSPM Farming

By

John Abraham Godson
& Esonu Nduka Udeala

Pilgrim Publication

Chapter 1

INTRODUCTION

Sweet Potato (*Ipomoea batatas*) is one of the major root and tuber crops of the tropic region of the world. Other root and tuber crops of this region include cassava (*Manihot esculenta* Crantz), yam (*Dioscorea* spp.), and edible aroids (*Colocasia* spp. and *Xanthosoma sagittifolium*). These are widely grown and consumed as subsistence staples in many parts of Africa, Latin America, the Pacific Islands and Asia.

Generally, Potatoes are low in calories. They are a good source of carbohydrate, Vitamin C and B6, Manganese, Phosphorous, Niacin, and Pantothenic acid and fiber which helps to lower total amount of cholesterol which could affect the heart. They are highly popular worldwide and prepared and served in a variety of ways. In recent times, potato is being used in baby-food formulation, cosmetics and pharmaceuticals and compounding animal feeds. The use of potatoes is growing every day.

Potatoes come in different types or varieties. Each type has its peculiar additional nutritional advantage. Some of the varieties are well known in some countries but unknown or unpopular in some other countries. This Farming Guide will focus on Orange-Fleshed Sweet Potato (OFSP) because of its peculiar advantages particularly needed to deal with malnutrition and food security.

Potato can thrive in a marginal soil and so can easily adopt to most environments. It grows best on light, sandy loams or on loamy sands which are moist, fertile and deep. So, it can be grown on land or in pots.

There are many literatures on potatoes. However, in this book, we are not just contributing to knowledge but providing practical guide to anyone who may want to grow potatoes for himself or for others to enjoy. Therefore, we will focus basically on the critical information for our goal.

According to FAO **International Year of the Potato 2008 Report**, *"Potato's role in the developing world's "nutrition transition" In many developing countries, and especially in urban areas, rising levels of income are driving a "nutrition transition" toward more energy-dense foods and prepared food products. As part of that transition, demand for potato is increasing. In South Africa, potato consumption has been growing in urban areas, while in rural areas maize is still the staple. In China, higher income and increased urbanization have led to increased demand for processed potatoes. Thus, the potato already plays a role in diet diversification in many countries. However, where other staple crops are available to meet energy requirements, potato should not replace them but rather supplement the diet with its vitamins and mineral content and high quality protein. Potatoes can be important staple foods, but balanced diets need to include other vegetables and whole grain foods."*[1]

It is this increased need for potato that has necessitated the involvement of many more people in potato farming.

Also, with rising number of people in the sub-Saharan Africa suffering from not only severe acute malnutrition but micronutrient malnutrition or 'hidden hunger', a problem of deficiency of essential micronutrients such as vitamin A, iron and zinc which unfortunately affects women of reproductive age, infants, and young children more than other segments of the society, it has become necessary to popularize the production and utilization of the

Orange-fleshed Sweet Potato (OFSP) as major staple crop because of its higher content of beta carotene (which is precursor for Vitamin A), iron and zinc. The idea is to help solve the malnutrition problem of weak immune system, visual impairment, night blindness in pregnant women and children, cognitive inability, retarded growth and reproductive potential, increased risk of disease which ultimately affects the general productivity of the populations; as well as guaranty food security in this part of the world.

We believe that the enthusiasm expressed in this work and the step – by – guide provided will stir the reader into doing something about Orange-fleshed sweet potato.

References

[1]Food and Agricultural Organization [FAO] **International Year of the Potato 2008 - Potatoes, Nutrition and Diet** p1, 2.

Chapter 2

BENEFITS OF ORANGE FLESHED SWEET POTATOES (OFSP)

Nutritional value of OFSP

One cup (200 grams) of baked sweet potato with skin provides:

1. Calories: 180
2. Carbohydrate: 41.4 grams
3. Protein: 4 grams
4. Fat: 0.3 grams
5. Fiber: 6.6 grams
6. Vitamin A: 769% of the Daily Value (DV)
7. Vitamin C: 65% of the DV
8. Manganese: 50% of the DV
9. Vitamin B6: 29% of the DV
10. Potassium: 27% of the DV
11. Pantothenic acid: 18% of the DV
12. Copper: 16% of the DV
13. Niacin: 15% of the DV

Orange fleshed sweet potato (OFSP) has several benefits and below are some of them.

Health Benefits

1. They are highly nutritious. This is why they are highly recommended for infants.
2. It is rich in vitamins and minerals: vitamin A, vitamin B-complex, vitamin C, beta-carotene, potassium, and calcium

3. Helps improve digestion due to its high fibre and Magnesium content. Being mainly starch they are simple to digest.

4. Helps Treat inflammation due to the presence of beta-carotene, Vitamin C and Magnesium. They are reliable cures for internal and external inflammation.

5. Helps Boost immune system due to abundance of beta-carotene which is a major anti-oxidant. Together with the minerals- phosphorus and iron, and the vitamins- C and B complex, they are excellent immune boosters.

6. Treats stomach ulcers- Due to the content of Vitamins B complex, C, beta-carotene, potassium and calcium, OFSP are very good in curing stomach ulcer. The fibre also helps in preventing constipation due to the appropriate movement of food. Constipation leads to acid buildup that causes ulcer.

7. Helps treat bronchitis and other respiratory ailments – The abundant presence of Vitamin C, iron and other nutrients serves as a good cure for bronchitis.

8. Reduces arthritis pain- due to the presence of beta-carotene, magnesium, zinc and Vitamin B. OFSP are useful in diminishing arthritis pain.

9. Helps control diabetes- due to the fact that they are categorized as low to high on the glycemic index scale, they release sugar into the bloodstream slowly unlike other starchy food. Thus, OFSP can be used to regulate blood sugar levels especially in people with diabetes.

10. Helps prevent dehydration- due to the sugar and fibre content, OFSP helps the body to retain water.

11. They have anti-oxidants and therefore may fight cancer. Sweet potatoes contain beta-carotene, a carotenoid that is essential in

protecting men from developing prostate cancer. In addition, extracts from sweet potatoes have been shown to accelerate the death of breast, gastric, colorectal and prostate cancers

12. Improves and supports healthy vision – due to the presence of Vitamins A, C and E which are important for a healthy vision. Deficiency of Vitamin A leads to night blindness, total blindness, as well as malformation of the retina. Deficiency in Vitamin C leads to the development of cataracts.

13. Can improve brain function, learning and memory due to the presence of anthocyanin which has antioxidant properties thereby making consumption of sweet potatoes memory enhancing.

14. Can help manage stress levels- due to the presence of magnesium. Magnesium has also been found to reduce insomnia in older individuals. Insomnia is a known trigger to stress, depression and anxiety.

15. Can help prevent cardiovascular disease – due to considerable amounts of Magnesium and potassium both of which are important in fighting hypertension and other cardiovascular diseases.

16. Contains antimicrobial properties. Some studies have shown that the ethanol and acetone extracts from sweet potato leaves possess antimicrobial properties against bacteria that cause pneumonia and typhoid. The fibre in sweet potato has also been shown to inhibit the growth of foodborne bacteria.

17. Can help improve hair and skin – due to high levels of Vitamins A, C, and E. These vitamins are also known to help in slowing aging.

18. Can boost fertility in women because of its content of iron which is essential in promoting fertility amongst women. Vitamin A is essential in improving reproductive performance in women of childbearing age.

19. Can help in weight management because the fibre can help in maintaining satiety thereby helping in weight management. The fuller you feel, the less you feel like snacking.

References

https://blog.agrihomegh.com/benefits-orange-fleshed-sweet-potato/

https://lacanadacarecenter.com/15-health-benefits-of-sweet-potatoes-according-to-science/

Chapter 3

PLANTING MATERIAL SELECTION AND PREPARATION

Potatoes have short gestation cycle of 3 to 4 months depending on the variety. There are two major ways of propagating potatoes. These are by vegetation – (vine cutting or seed tuber), and by seed. Propagation by seed is mainly done in research institutions through which genetic manipulation can be carried out. However the variety of potato also determines which method of propagation is best suitable. So the first thing is to choose what type/variety of potato you want to plant.

Selection of Planting Material

There are many domesticated potato varieties and thousands more in the wild. The ones that are mostly eaten include:

1. Sweet potato
2. Irish Potato
3. Purple Skin (Orange-fleshed) Sweet potato
4. Orange-fleshed Sweet Potato
5. Purple-fleshed Sweet Potato

POTATO VARIETIES AND THEIR NUTRITIONAL VALUES

Sweet Potato

Sweet Potato has white or cream colour skin and white/cream flesh. It has good quantity of carbohydrates among other nutrients, and according to its name, it is sweet and can be used as sweetener especially in preparing drinks like kunu, salad, etc.

It is propagated both by vine cutting and seed tubers.

Irish Potatoes

Irish potato has yellow/deep cream colour skin and flesh. It can also come in red colour. In addition to the other nutrients, Irish potato has Folate which is an essential nutrient needed by the body but cannot be stored in the body. So it can only be gotten by daily consumption of foods rich in it.

It is propagated mainly by seed tubers (potatoes).

Purple Skin (Orange-fleshed) Sweet Potato

king J

This is a type of Orange-fleshed Sweet Potato (OFSP). However its skin is light Purple while the flesh is light orange.

It has Vitamin A and rich in antioxidants that protect the body from free radicals. It helps to protect the body from chronic illnesses like cancer, heart disease and aging. It has higher dry matter (36%) than the deep orange coloured OFSP which has 22% dry matter.

It is propagated by vine and seed tubers.

Orange-fleshed Sweet Potato Mothers' Delight

Orange-fleshed Sweet Potato (OFSP) has orange colour skin as well as deep orange flesh. It is high in beta-carotene which is converted to Vitamin A in the body. Vitamin A is key for maintaining healthy skin, vision and organ function. It is also rich in antioxidants that protect the body from free radicals. It helps to protect the body from chronic illnesses like cancer, heart disease and aging. The leaves are used as vegetable for cooking and is very rich in Protein. The leaves are also used in compounding animal feeds.

Propagation is by vine cutting and seed tubers.

Purple -fleshed Sweet Potato

Purple-fleshed sweet potato has deep purple skin as well as purple flesh. It is also rich in antioxidants that protect the body from free radicals. It helps to protect the body from chronic illnesses like cancer, heart disease and aging. It is highly needed in pharmaceutical and cosmetic industries.

Propagation is by seed tubers.

The best potato varieties are those that are preferred by consumers in your community or the available off-taker.

Our variety of choice in this Guide is orange-fleshed sweet potato. It is propagated by vine cutting and by seed tuber. Though it also grows flowers but cross pollination which eventually results in seed (fruit seed) hardly takes place in the farmer's field before harvest. Even though in the research institute this is the major way of getting new verified seed. For a large-scale or commercial farming of OFSP vine cutting is a more recommended planting material.

Select healthy planting material (Vine/Seed tuber) free of disease or pest infestation. Never use dried materials as its viability has been compromised. It may not sprout and if it eventually sprouts, its yield will be poor.

Diseased
OFSP Vine stem.

The potato vine/tuber has short life span once it is removed from the ground. It either dries up or rots away. Therefore you need to have access to fresh and healthy planting material - vine/tuber. The vine cutting from stem apex sprouts faster and yields better than the ones from the base of the vine and so is the recommended vine cuts to use.

Preparation of Planting Material

*OFSP Vine stems growing
in the garden*

*OFSP Vine stem being
cut out for use*

*OFSP Vine stem being
inspected*

*OFSP Vine stem cut
for planting*

From fresh crawling vine stem, cut 10 to 30 cm or 4 – 6 nodes for planting. This measurement should be maintained along the vine stem except when nearing the tip of the vine in which case the last cut should be given not beyond the brown/purple colour area of the stem. This is to ensure that the last cut sprouts. If the nodes on the vine are far apart like the ones in the above pictures, it will be necessary to take your cut 3 to 5 cm to the node. This is because during planting, you will plant the vine cut slanting bury at least two nodes into the ground at about 10 – 15 cm.

You will have to cut the quantity of vines you need for your farm land before planting.

However, if you are getting pre-cut vines you will need the required number of bundles as you go for planting.

Some OFSP Vine cuts
being preserved for plant
under a shade

OFSP Vine cuts
pre-sprouted before planting

If your vine cuttings are supplied some days before you will plant them, and the weather is dry, you will need to keep them from drying up by placing them on a wet ground under a shade. It is advisable to make sure that the end that is to the buried is the side on the ground as it will start developing roots within two to three days   as long as you keep them wet every day.

*OFSP seed tubers
for planting*

*OFSP seed tuber
that is sprouting*

*OFSP seed tuber
that has sprouted
ready for planting*

If you are planting with seed tubers, make sure they are fresh and healthy and not dried or diseased. Pictures above show some OFSP seed tubers for planting. You can pre-sprout before planting by keeping them in damp warm place for 3 – 5 days. They will begin to develop the sprout/eye so that you will know which side to bury and which side to face up while planting.

In the next Chapter, we will deal with planting operations in full.

Chapter 4

LAND PREPARATION

Sweet potatoes can be grown in almost all types of soil. The best soil however is sandy loam with adequate amounts of organic matter. Sweet potato grows best and produces smooth, well-shaped storage roots in a well-prepared soil. Light-textured soils generally produce sweet potato roots with smoother skins than do heavy-textured soils. The lighter soils may produce longer roots, however.

The land should be cleared and all organic debris can be cultivated back into the soil. Burning of organic debris should be avoided. Burning generally kills soil microorganisms. OFSP being a root crop, needs that the land be well prepared to allow for good aeration, adequate moisture availability and drainage as well as appropriate microbial activity. The land should be well ploughed to at least 20cm deep and harrowed several times to break up any impediments to root development. OFSP can be planted in beds, mounds or ridges but ridges are recommended in commercial OFSP production- especially in places with high soil moisture. Ridges help in drainage control.

Soil should be cultivated when damp but not too wet. Cultivation of dry or very wet soil can break up the soil structure, leading to poor drainage and aeration, surface crusting, cracking allowing entry of weevils, and greater susceptibility to erosion.

The following management practices could be included during land preparation to ensure maximum yield.

Organic matter incorporation – Organic manure (Poultry manure, Pig manure,

Goat or cow dung, Compost) can be incorporated and ploughed into the soil. Doing this fulfills the following functions:

1.   Organic manure is a source of the needed nutrient for plant growth and tuber formation.

2.   It also provides a good environment for microorganisms thrive and promote overall development of the crop

3.   Organic matter also helps maintain a good soil structure that allows for adequate soil aeration and moisture retention thereby aiding good root growth and excellent tuber formation.

Care should however be taken not to overdo this as too much nitrogen will promote growth of leafy vegetation and not necessarily tuber formation and development.

Sweet potatoes do well in slightly acidic soil unlike most other vegetables. A soil pH between 5.6 and 6.5 is ideal. Most soils in the tropics and subtropics are rather acidic and therefore ideal for OFSP.

Ridges should be between 30cm-45cm high and about 90cm – 120cm apart measuring the highest points of the ridges. The use of heavy machinery, such as tractors, for soil preparation may compact the sub-soil producing a hard pan or compaction layer. This problem is greater on clay soils than on sandy soils. A hard pan can impede drainage, causing waterlogging, poor development of beneficial organisms, and poor root growth, which consequently leads to poor plant growth and poor-quality roots development.

A Picture of Simdozi International Ltd OFSP Farm in
Abuja Nigeria, April 2020

References:

Land preparation and cultivation by Vilma Amante and Jane O'Sullivan
https://keys.lucidcentral.org/keys/sweetpotato/key/Sweetpotato%20Diagn
otes/Media/Html/TheCrop/CropManagement/Crop%20mgt%20Land%20pr
ep.htm

What Should the Ground pH Balance Be for Growing Sweet Potatoes?
https://homeguides.sfgate.com/should-ground-ph-balance-growing-sweet-
potatoes-71311.html

Sweet Potato agronomy by Maniyam Nedunchezhiyan, Gangadharan Byju and Susantha K Jata

http://www.globalsciencebooks.info/Online/GSBOnline/images/2012/FVCSB_6(SI1)/FVCSB_6(SI1)1-10o.pdf

Sweet potato – Commercial vegetable production
https://athenaeum.libs.uga.edu/bitstream/handle/10724/12300/b677.pdf?sequence=1

Chapter 5

PLANTING OPERATIONS

Having done pre-planting activities including site selection, land preparation, seed (planting material) selection and preparation, we are now set to go into the farm and plant the materials we have earlier prepared.

OFSP needs good supply of water to sprout, survive and do well. It is therefore advisable to plant your orange-fleshed sweet potato vine cuttings or seed tubers once there is reasonable water in the soil. The plant needs soft wet (not flooded) soil within its first six weeks as this is the period it develops roots and bulks into tubers. For those in the tropical region, you are encouraged to plant early in the rainy season. Doing this will guaranty two cycles of production before the cessation of rainfall. On the other hand, if there is availability of irrigation facility, OFSP can be planted any time of the year as long as water is made available constantly at the most critical times in its cycle.

Pre-sprouted vine cut ready for planting

OFSP Vine Cutting being planted on a ridge.

Some ridges planted with OFSP Vine Cuttings

Your orange fleshed sweet potato can be planted on ridges, mounds or even pots. In whatever medium you plant, the vine must be planted slanting with at least two nodes buried about 7-14 cm in the soil while the remaining nodes on the cut are above the soil. This makes it easy for sprouting.

If the planting material is seed tuber, it should be buried 5-10 cm slanting with the eye or sprout facing upward.

If planting on ridges, the planting distance is between 25cm and 30cm along the ridge.

After two weeks of planting all viable vines/tubers must have sprouted. If there are those that did not sprout, you can re-supply at this time so as to still have relatedly uniform harvest at the end.

Chapter 6

POST PLANTING OPERATIONS

There are various operations that need to be done before and after the planting operation (pre-planting and post-planting operations). Post-Planting operations include Weeding, Fertilizer application, Pest control, harvesting and storage. We are going to talk about weeding and Fertilizer application in this chapter while the other operations will be dealt with in subsequent separate chapters.

Weeding

Weeding is a very important farm operation as weeds can compete with crops for soil nutrients, soil moisture, air, light and space. Weeding can be done both as a pre-planting operation and as post-planting operation. Good weed control starts from a proper land preparation, ploughing and harrowing. Weeding can be done manually or by the use of herbicides.

Manual weeding is the most desirable especially of the leafy vegetation would be part of the products to be consumed by humans or livestock. Although it may be more time consuming and labor intensive – it is safer and more ecological. When the OFSP are planted closely and when weeding is done early enough, OFSP develops adequate foliage to cover the farm land preventing more weeds to emerge

Use of herbicides can be done as a pre-planting operation and as a post-planting operation.

21

Pre-planting chemical wee d control in sweet potato.

Active Ingredient	(Trade name) amount of product	Weeds controlled/remarks
Carfentrazone Up to 0.031	(Aim®) 2 EC or 1.9 EW Up to 2 fl. oz.	Apply as a preplant burndown for emerg ed broadleaf weeds. Use crop oil concentrate or nonionic surfactant at recommended rates. Maximum rate of 0.096 lb. a.i./A per season. No pretransplant interval.
Clomazone 0.49 –0.75	(Command®) 3 ME 1.3 – 2.0 pt.	Annual broadleaf and gr ass weeds. Use lowe r rates on coarse soils. Apply within 5 days of transplanting.
DCPA 4.5–6	(Dacthal®) W -75 6–8 lb. (Dacthal®) 6 F 6–8 pt.	Annual broadleaf and grass weeds. Apply immediately after transplanting. May be applied as a layby later in the season for pre -emergence control.
Flumioxazin 0.096	(Valor®) 51 WDG 3 oz.	Annual broadleaf weeds. Do not use transplants that were harvested 2 days before application. Severe injury occurs if applied after transplanting. Apply 2–5 days before transplant and minimize soil disturbance after application.
Glyphosate	(Various formulations) Consult label	Emerged broadleaf and grass weeds. Apply as a preplant burndown. Consult label for individual product directions.

Napropamide 1.0–2.0 (Devrinol® DF-XT) 50 DF 2–4 lb. — Annual broadleaf and grass weeds. Apply immediately after transplanting. If rainfall does not occur within 24 hours after application, incorporate or irrigate 2–4 inches deep.

Pelargonic acid (Scythe®) 4.2 EC 3–10% v/v — Emerged broadleaf and grass weeds. Apply as a preplant burndown treatment. Product is a contact, nonselective, foliarapplied herbicide with no residual control. May be tank mixed with soil residual compounds.

Pyraflufen 0.001–0.003 (ET Herbicide) 0.208EC — Emerged broadleaf and grass weeds. Apply as a pre-plant burndown.

Post-planting chemical weed control in sweet potato

Active Ingredient	(Trade name) amount of product/A	Weeds controlled/remarks
Carfentrazone Up to 0.031	(Aim®) 2 EC or 1.9 EW Up to 2 oz.	Emerged broadleaf weeds. App ly as hooded application to row middles only. Use crop oil concentrate (COC) or nonionic surfactant (NIS) at recommended rates. Contact with the leaves will cause injury. PHI 0 days.
Clethodim 0.09–0.25 0.07–0.25	(Select®, Arrow®) 2 EC 6–16 fl. oz. (Select Max®) 1 EC 9–32 fl. oz.	Perennial and annual grass weeds. Use higher rates under heavy grass pressure or larger grass weeds. Do not apply more than 0.5 lb. a.i./A. Consult label for required adjuvant. PHI 30 days.

Fertilizer with potassium and phosphate levels that are higher than nitrogen levels should be chosen for tuber crop production. Because potatoes are root crops that grows below the surface of the soil, phosphate and potassium are more beneficial to potato growth. A fertilizer formula of 5-10-10 or 8-24-24 is ideal for sweet potatoes.

For production where tuber crops and vine cuttings are goals of the operation – fertilizer formula of 15:15:15 is appropriate. For an acre of land – 150kg (3 bags) of NPK fertilizer is recommended.

Generally, 2-3 applications are sufficient

1.	First application: 4-6 weeks

2.	Second application: 10-12 weeks

3.	Not later than 3 weeks before harvesting

Water-soluble product for fertilizing sweet potatoes are preferable. Granular fertilizer is fine while the plants are still small. But they can be burned if it comes in contact with the plants. As the plants grow and start to spread out, it is much easier to apply a fertilizer you can mix into water. In addition, water-soluble fertilizer can be used during regular irrigation (if in use), which is less time-consuming. If using the granular fertilizers, they should be applied by the sides of the ridges or beds.

References
Sweet Potato agronomy by Maniyam Nedunchezhiyan, Gangadharan Byju and Susantha K Jata
http://www.globalsciencebooks.info/Online/GSBOnline/images/2012/FVCSB_6(SI1)/FVCSB_6(SI1)1-10o.pdf.

Weed management in sweet potato by Peter Dittmar and Nathan S. Boyd

Fluazifop 0.1–0.25	(Fusilade® DX) 2 EC 6–16 fl. oz.	Perennial and annual grass weeds. Include an NIS or COC in the spray solution. PHI 55 days.
Pelargonic acid	(Scythe®) 4.2 EC 3–10% v/v	Emerged broadleaf and grass weeds. Apply as a hooded application to row middles only. Contact with the leaves will cause injury.
S-metolachlor	(Dual Magnum) 7.62 EC	Grass, broadleaf, and nutsedge. Label is a Third - Party Registration (TPR, Inc.). Use without a signed authorization and waiver of lia bility is a misuse of the product.
Sethoxydim 0.19–0.47	(Poast®) 1.5 EC 1.0–2.5 pt.	Grass weeds. Maximum use of 5.0 pt./A applied per season. Include a COC. Unsatisfactory results may occur if applied to grasses under stress. PHI 30 days.

Fertilizer Application

The amount of manure or fertilizer applied should be determined by the goal of the farm. If the goal of the farm operation is to produce planting materials – vine cuttings, vegetational growth becomes of prime importance. More manure and nitrogen fertilizer is needed to promote a more robust vegetational growth. If tuber production is the goal, less of such is needed. As mentioned earlier – Organic manure is very good for OFSP growth and development as it tends to make use of soil nutrients heavily. 2-4 tons of organic manure per acre should be ploughed to the ground during land preparation.

https://edis.ifas.ufl.edu/pdffiles/WG/WG03900.pdf

Watering and fertilizing Sweet potatoes https://www.backyard-vegetable-gardening.com/fertilizing-sweet-potatoes.html

Sweet potatoes (Ipomoea Batatas L.) production
https://www.nda.agric.za/docs/Brochures/PG_SweetPotato.pdf

Fluazifop 0.1–0.25 (Fusilade® DX) 2 EC 6–16 Perennial and annual grass weeds. Include an NIS fl. oz. or COC in the spray solution. PHI 55 days.

Pelargonic acid (Scythe®) 4.2 EC 3–10% Emerged broadleaf and grass weeds. Apply as a v/v hooded application to row middles only. Contact with the leaves will cause injury.

S-metolachlor (Dual Magnum) 7.62 EC Grass, broadleaf, and nutsedge. Label is a Third - Party Registration (TPR, Inc.). Use without a signed authorization and waiver of lia bility is a misuse of the product.

Sethoxydim 0.19–0.47 (Poast®) 1.5 EC 1.0–2.5 Grass weeds. Maximum use of 5.0 pt./A applied per pt. season. Include a COC. Unsatisfactory results may occur if applied to grasses under stress. PHI 30 days.

Fertilizer Application

The amount of manure or fertilizer applied should be determined by the goal of the farm. If the goal of the farm operation is to produce planting materials – vine cuttings, vegetational growth becomes of prime importance. More manure and nitrogen fertilizer is needed to promote a more robust vegetational growth. If tuber production is the goal, less of such is needed. As mentioned earlier – Organic manure is very good for OFSP growth and development as it tends to make use of soil nutrients heavily. 2-4 tons of organic manure per acre should be ploughed to the ground during land preparation.

Chapter 7

PESTS AND PATHOGENS

The essence of this book is to increase the awareness and simplify the knowledge of potato (OFSP) production. However, it is an obvious fact that potato production is faced with some challenges including those of pest and pathogens. There are some pests and pathogens that affect the potato crops in general and orange-fleshed sweet potato in particular. In this chapter, we will discuss some of them and how to handle their attack on your farm to ensure that your labour is not in vain.

Some of the potato pests are insects which have different life stages (metamorphosis) that can sometimes be confusing to a farmer because of their unrelated physical features. The farmer needs to know what is affecting his farm at every point in time so as to be able to avoid a more serious damage in future. Examples of these insects include: **Sweetpotato Weevil (Cylas spp.), Rough Weevil (Blosyrus sp.), Clear Wing Moth, Sweetpotato Butterfly, Sweetpotato**

Hornworm and Armyworms. They grow from egg to larva, to pupa and finally to adult and behave differently at the different stages. They attack mostly the root/tuber. Whereas the adult insect is dangerous, the larvae cause large-scale infestation by chewing their way through your roots/tubers. As a conscious farmer, when you notice a few adult insects in your farm, the earlier you get rid of them the better because most insect populations grow very quickly.   They lay their eggs and before you know it the whole farm is already captured by these harvest destroyers.

During land preparation you may notice a lot of pupa or larvae in the soil. Do not take this for granted. If you do nothing at this stage and go ahead to plant, the young tubers that will grow out of your planted vine cuttings would be ready food to these awaiting armies of pests. So, eliminate them directly or by applying necessary pesticides/insecticides if they are wide spread. A healthier human health method is to allow other members of the eco system which are hostile to these pests to kill them.   Such pests' attackers which are not dangerous to your potato are animals or insects that hunt, kill and eat other creatures.   They include spiders, ants, ground beetles, earwigs, ladybird beetles, lacewings, and flower beetles.

The other type of insect pests have more recognizable lifecycle development because they go through incomplete metamorphosis. They include **Aphids, whitefly, Grasshoppers and Mirids (Sucking bugs)**. While Aphids and Whitefly spread sweet potato virus, Grasshoppers (may sometimes be considered a minor pest) eat up the potato leaves and if they are in a reasonable number, can impact on the yield of the farm. On the other hand, Mirids and other Sucking bugs typically feed on the young shoots and leaves causing black lesions and leaf puckering. This can hamper the growth of the plant and eventually the yield.

According to FAO International Year of Potato 2008 Report on **Potato Pest and Disease Management** p1

1.	*Colorado potato beetle (Leptinotarsa decemlineata) is a serious pest with strong resistance to insecticides.*

2.	*Potato tuber moth, most commonly Phthorimaea operculella, is the most damaging pest of planted and stored potatoes in warm, dry areas.*

3.	*Leafminer fly (Liriomyza huidobrensis) is a South American native common in areas where insecticides are used intensively.*

4.   *Cyst nematodes (Globodera pallida and G. rostochiensis) are serious soil pests in temperate regions, the Andes, and other highland areas.*

Viruses are disseminated in tubers and can cut yields by 50 percent.

1.   *Late blight, the most serious potato disease worldwide, is caused by a water mould, Phytophthora infestans, that destroys leaves, stems, and tubers.*

2.   *Bacterial wilt, caused by the bacterial pathogen, leads to severe losses in tropical, subtropical, and temperate regions.*

3.   *Potato blackleg, a bacterial infection, causes tubers to rot in the ground and in storage.*

The dangers of Pests Infestation

Generally, sweet potato has a short gestation period: 90 – 120 days from planting to harvest. The pests that attack it have short but fast development stages. This can lead to pest population build up which will destroy any effort. For instance, it takes a sweetpotato weevil one month to grow from egg to adult. When one weevil lays about 100 eggs which in a month matures to 100 adults of which 50 of them are females, you discover that by the second month of your crop, 50 female adult weevils laying 100 eggs each would produce 5,000 adult weevils which will be a great disaster in the farm. Meanwhile the other 50 male adults are busy doing their own mischief. That is why it is very important to deal with these on time before they get out of hand.

Some factors affect the development of these insects. They are environmental

Sweet Potato weevil

Rough weevil

Sweet Potato butterfly

and nutrition. Insects develop faster in warmer temperatures, but would slow down development or even die when the temperature gets really hot. On the other hand, the type of food available for the insects at different stages of their lives affect how they develop or behave. For instance, some insects, can live and feed on an alternative host plant, which is plant other than their main target waiting for the emergence or maturity of their target plants at which point they rapidly spread. For example, the water spinach plant which sometimes grows in between sweet potato crops is a ready alternative host to sweetpotato weevil.   It is therefore necessary to pay attention to weed management particularly within the second and fourth week of your OFSP crop.

Whereas some pests and diseases are within an environment others are transported via planting materials, movement of wind, or even by the farmer via his dress/footwears.

Management of Pests and Pathogens

The current trend in plant pest and disease management is Integrated Pest Management (IPM) which is a combination of options and tactics with the purpose of reducing crop losses by pest, preventing distortion/destruction of eco system by use of chemicals, and minimizing risks to human health. It involves careful study of the environment, understanding the agricultural practices of the local community, monitoring the crop sanitation and pest biology. The farmer pays more attention to the health of his planting materials and his own activities to ensure that he does not transfer diseases/pests from one farm to another. He ensures that he uses crop varieties that are more resistant to the prevalent pests/diseases in his community.

Clear wing moth

Sweet potato horn worm

Army Worm

References:

Jenny Lazebnik: **Potatoes, Pathogens and Pests: Effects of genetic modification for plant resistance on non-target Arthropods**. PhD thesis, Wageningen University, Wageningen, the Netherlands (2017) With references, with summary in English. http://dx.doi.org/10.18174/411738

Food and Agricultural Organization (FAO) **International Year of Potato 2008 Report on Potato Pest and Disease Management p1**

Everything You Ever Wanted to Know about Sweetpotato. Topic 7 - Sweetpotato Pest and Disease Management Reaching Agents of Change ToT Training Manual   © International Potato Center, Lima, Perú, 2018 **Orange-Fleshed Sweetpotato: Production, Processing and Utilization A Community Training Manual,** Helen Keller International, Nigeria and International Potato Centre, Nigeria ,February, 2015

Chapter 8

HARVESTING AND STORAGE

OFSP Farm ready for harvest

As the days go by your OFSP crop is getting ready for harvest. All other things being equal, that is if there was sufficient water supply for the crops at least in the first 4- 6 weeks, by the end of the 12th week (3 months) they are ready for harvest. However, if there was not much rain or water supply at this critical time but it came later, then it is advisable to allow the crop get to the 4th month before harvest. Other indicators showing that the tubers are mature include the breaking of the ground around the tubers, some of the leaves turning yellowish and drying up, sometimes the top/shoulder of the tuber may be bulging out from the ground. Before harvesting the entire field, do a sample harvest of a few stands to know the state of the farm. It must be mentioned that any tuber bulging out of the ground should be covered with soil to avoid exposing it to diseases and rodents. If matured tubers are left to keep growing in the ground beyond 4 months they begin to fiberize and lose their taste and crunchiness.

36

When to harvest

Harvested OFSP Tubers

Orange-fleshed sweet potato has a short shelf life but very tasty while fresh. It also has very brilliant orange skin colour while fresh and so very appealing to a buyer. Therefore, if you farm for business – to sell and make money, it is preferred that harvest be done when you have ready buyers so that they can take delivery while your tubers are fresh. So, if your farm displays the above signs of readiness for harvest and you have a ready market, you should go on with the harvest.

Implements for harvest

You need the correct implements to do a good harvest. The tubers are to be removed from the ground with much care so as to minimize bruises and cuts on the tubers because these will reduce the marketability of the tubers as well as increase their vulnerability to rotting and diseases during storage. The right tools to use for harvest are digging rods/forks, hoes, and shovels.

Post-Harvest Activities And Storage

Harvested OFSP Tubers Being Washed

Harvested OFSP Tubers should be washed in water especially if harvested during rainy season when the ground is wet and the roots come out with muds.

The washed tubers should be dried and weighed before sorting, packaging or storage. This will help you keep a record of the weight of the harvest from your farm for the season.

Not all the tubers are big enough for selling. Therefore, sort out the sellable sizes and weigh on a scale. This helps you know the actual quantity/weight you will sell.

If you selling to a standardized off-taker, the tubers are normally sold in weights/kilogram. However, if you are selling to non-standardized buyer, the tubers may not be sold in weights.

Washed OFSP Tubers Being Dried

If the tubers are to be stored for future use, it is recommended that they should be cured before storing.

Curing is a wound healing process and prevents rot during storage. To cure the tubers, spread the harvested roots in a big ventilated room under room temperature for 4-7 days before storage. Thereafter, dust the tubers with wood ash and store. Only clean, unbruised and disease-free tubers should be stored.

Storage of OFSP Tubers

Harvested OFSP tuber has short shelf life of 3 to 5 months. The tubers should be cured before storing. The cured tubers can be stored in the following places:

1.   In Pit under-laid with dry grasses/leaves

2.   In Saw dust heap

3.   On Rack Storage

4.   In Dry sand under ventilated room with room temperature.

However, the unharvested tubers can be stored in the ground for future harvest and use. This is done by cutting the vines off matured tubers and cover the ground well without harvesting the tubers. This can store for another 3-4 months without fiberizing at harvest.

References:

Everything You Ever Wanted to Know about Sweetpotato. Topic 7 - Sweetpotato Pest and Disease Management Reaching Agents of Change ToT Training Manual   © International Potato Center, Lima, Perú, 2018 Orange-Fleshed Sweetpotato: Production, Processing and Utilization A Community Training Manual, Helen Keller International, Nigeria and International Potato Centre, Nigeria ,February, 2015

Chapter 9

ECONOMICS OF ORANGE-FLESHED SWEET POTATO

Orange-fleshed sweet potato has great economic value and opportunities. The Orange-fleshed sweet potato value chain is a long, lucrative and developing one. Every actor in the chain is a winner. As OFSP is now the best and preferred variety of potato due to its nutritional content and colour, it is daily receiving acceptability across the globe. Anyone involved in the production, processing, marketing or even consumption of OFSP has much to gain. In this Chapter we will look at the economics of OFSP under the following subheads: OFSP for food, OFSP as raw material, OFSP and other businesses.

OFSP FOR FOOD

Orange-fleshed sweet potato is rich in vitamins and other micro nutrients and so is much needed in fighting malnutrition and other diseases and food insecurity. *OFSP tubers have been utilized as food in their fresh form after cooking, as flour and in the grated and mashed (commonly known as puree) forms. OFSP flour has been used locally at domestic level and in industrial production of bakery products. Some of the bakery products in which OFSP flour is incorporated as an ingredient are cakes, bread, muffins and buns. Mashed OFSP has also been used for flour substitution in Golden bread.*[1]

In his paper on **Combatting Poverty With the Orange-fleshed Sweet Potato** *Dr Petros Nyakunu of PORENet Mozambique wrote: "OFSP mixes very well as an ingredient in traditional recipes, as well as in more processed products. You can be creative and try it steamed, boiled, mashed, fried, roasted, baked or use it to*

41

make products for weaning or as preserves. OFSP products can provide those who buy them with a varied diet and at the same time generate income for the producers. Growing OFSP can also help empower women, enabling them to give their families food that will keep them healthy, and providing them with a new source of money for their other needs."[2]

In the countries where OFSP is well known, the tubers are sold at a premium price higher than other varieties of potatoes because of its higher nutritional value. This places the OFSP farmers at advantage.

Dishes from OFSP are great delicacies in top hotels and eateries; just as the tubers are in demands in popular chain stores where expatriates easily access them. Meeting the all-year-round supply gap for OFSP is a huge business opportunity.

Even though the OFSP tuber may have relatively short shelf life, it can be processed into longer lasting forms and still retain the nutrition values. Therefore, the OFSP farmers can always be sure of ready market for their produce.

In the next Chapter, we will discuss in more details the various dishes and products that can be prepared from OFSP.

OFSP AS RAW MATERIAL

"In industrial production, OFSP has been used to produce products such as chips, crisps, flour, puree, juice, bread and other bakery products. In Asian countries, sweetpotato pickles and cubes are produced commercially and are known for their β-carotene rich property. Some of

the OFSP roots such as Beta 1 and Beta 2, main varieties grown in Indonesia, are high in moisture thus are not consumed directly as roots but in derivative products In such cases, OFSP roots are processed and serve as functional ingredients. Production and use of OFSP puree as functional ingredients in food processing has been done for over three decades in the United States (US)."[3]

In the developing countries, there are few industries that are into industrial processing of OFSP. Even large-scale production of OFSP food products is still open for investment.

Potatoes are used in many of the biggest industries in the world, including wood, paper, textile and oil. Potato is used for starch/adhensive production which is needed in woods, textile and paper industries for gumming wood/papers or stiffening cloths.

OFSP vines/leaves are used in formulating animal feeds. So, production of vines could be for this purpose and the farmer will make his money both from the vines as well as the tubers that will be harvested at the end.

OFSP AND OTHER BUSINESSES

Preservation of OFSP vines from one farming season to another is pretty hard particularly in tropical region where this crop is grown. Farmers hardly have their own vines at the beginning of a new season. Orange-fleshed sweet potato vines are therefore not easily available. So, farmers involved in vine multiplication can make double income by first growing vines for sale, and eventually harvesting the tubers thereafter.

PROCESSING AND VALUE ADDITION FOR OFSP

OFSP tubers can be processed into different food forms, like garri, flour, chips, grits, juice, etc. This reduces the bulkiness of OFSP as a primary agricultural product, brings value addition and makes it more available to many more people. Below are some of the value addition products.

Fresh OFSP Tubers

Fresh OFSP Tubers Slices

OFSP Tubers Slices being Fried

Fried OFSP Chips

Commercial Orange Fleshed Sweet Potato Farming

This can be preserved and served as snacks at desired time. It is a value addition that makes OFSP product easy for packaging and distribution to even places where OPSP is not produced.

Fresh OFSP tubers sliced and sun dried

This can be used to prepare different meals particularly weaning meals for babies and children.

Dried Sliced OFSP ground into powder

TABLE 2
COST OF LAND AND LABOUR INPUTS

S/NO	DESCRIPTION/ACTIVITY	No. of Man days	Cost/ Man Day (₦)	Quantity Reqd	Unit Cost (₦)	Total cost
1	Lease of land (1 Acre)			1	15,000	15,000
2	Land Clearing					
	Cutting of bush (per Ha)	2	3000			6,000
	Stumping (per Ha)	2	2500			5,000
3	Land Preparation					
	Ridging	6	3500			21,000
4	Manure/Fertilizer Application	4	1500			6,000
5	Planting	5	2000			10,000
6	Application of Herbicide	2	1,500			3,000
7	Weeding	4	2000			8,000
8	Application of Pesticide (where applicable)	2	1,500			3,000
9	Harvesting					
	Cutting of vines/packaging	6	2,000			12,000
	Harvesting of Tubers	6	2,500			15,000
	Packing of Tubers	4	1,500			6,000
	Cost of water				2000	2,000
	Cleaning of Tubers	4	1,500			6,000
	Sorting	4	1,500			6,000
10	Marketing					
	Packaging	5	2,000			10,000
	Transportation	Lump			25000	25,000
	Total					159,000

ORANGE-FLESHED SWEET POTATO PRODUCTION BUSINESS

The objective of this guide is to raise men and women who will develop interest in the production of the OFSP. Increased production of this crop particularly in the sub-Saharan Africa will guaranty the availability of the crop for use as staple and in dealing with food insecurity. We have therefore included production business calculations to show the profitability or otherwise of the enterprise. Though the OFSP value chain is long, we only concentrated on the production aspect as we cannot demonstrate every aspect of the value chain. We believe that the calculations below will be an encouragement for someone thing of investing in this industry.

ESTIMATED COST BENEFIT ANALYSIS OF ORANGE-FLESHED SWEET POTATO PRODUCTION ON ONE ACRE

TABLE 1
AGRIC INPUTS

S/NO	DESCRIPTION	QUANTITY REQD	UNIT COST (Naira)	TOTAL COST
1	Vine Cuttings	222* bundles	700	155,400
2	Manure/poultry droppings	2000kg	4000/mt	8,000
3	Fertilizer (NPK 15:15:15)	150kg (3bags)	6000/bag	18,000
4	Chemicals:			
	Herbicide	1 litres	6000	6,000
	Insecticides/pesticide	1 litres	6000	6,000
	Total			**193, 400**

Source: Simdozi International Ltd Farm Records 2020

Note: * Plant population of 11,111 per acre.

Source: Simdozi International Ltd Farm Records 2020

Total Cost of Production: N193,400 + N159,000 = N352,400

Table 3
PRODUCTION OUTPUT

S/NO	DESCRIPTION	QUANTITY (KG/BUNDLES)
1	Tuber	3,155 kg (3.15mt)
2	Vine Cuttings	740 bundles
3	Farm Residual (for animal feed)	Bulk (entire farm)

Source: Simdozi International Ltd Farm Records 2020

Table 4
SALES OUTPUT

S/NO	DESCRIPTION	QUANTITY KG/BUNDLES	UNIT PRICE (Naira)	Total (Naira)
1	Tuber	3,155 kg (3.15mt)	150	473,250
2	Vine Cuttings	740 bundles	500	370,000
3	Farm Residual	Bulk (entire farm)	15000	15,000
TOTAL INCOM E				**858,250**

Source: Simdozi International Ltd Farm Records 2020

Total Revenue/Income = N858,250

density of 11,111/acre. To maximize the land lease for a year, the farmer is better off by doing two season farming. Also, he can increase the plant density to 14,666/acre or even 22,222/acre by planting at 30cm spacing or 20cm spacing instead of 40cm spacing used for this calculation. The only thing is that as the inputs go up the cost of production increases; which directly yields higher output and higher profit at the end. The marginal difference is quite remarkable.

This calculation is purely for production. Processors and marketers still have so much to benefit from OFSP value chain. The business is even more profitable when it is vine multiplication.

OFSP VINE MULTIPLICATION AS A BUSINESS

Though orange-fleshed sweet potato has gained acceptance in some Sub Saharan African countries, it is still a new crop in many more. The planting vine is also easily lost to dryness after each cropping season. Though propagation can be done via seed tubers but economy of scale can only be secured through vine propagation.

Also, in most places OFSP is planted in May/June and harvested around September. Hardly do people go into second cycle via irrigation. Therefore by the time it is season for OFSP planting in the following year, there is hardly any planting vines to use. This situation creates a good business opportunity for anyone who goes into vine multiplication as his major focus. The next attraction for vine multiplication is that it requires very minimal land space and agricultural inputs. Yet, the output is overwhelmingly enticing.

The other advantages of rapid vine multiplication cannot be over emphasized. It guarantees availability of planting materials all year round, thereby ensuring

Profit/Loss = Total Revenue/Income – Total Cost = 858,250 - 352,400 = 506,000

Total Net Benefit = N506,000

Using Gross Margin Analysis to evaluate the level of profitability:

Gross Margin Analysis = Net Profit/Total Cost x 100

= 506,000/352,400 x 100 = 1.43 or 143.58%

Using Benefit Cost Ratio (BCR) to check the viability of the business:

The Benefit Cost Ratio (BCR) is given as: BCR = $\sum$B/ $\sum$C Where $\sum$B = Total net benefit and $\sum$C = Total cost of production of the enterprise.

BCR = Total net benefit/ Total cost of production of the enterprise.

= 506000/352,400

= 1.43

Note: The methodology of interpreting benefit cost ratio indicates that BCR greater than 1, means that the Net Profit Value (NPV) of the project benefits outweigh the Net Profit (NPV) Value of the costs. Therefore, the project should be considered if the value is significantly greater than 1. If the BCR is equal to 1, the ratio indicates that the NPV of expected profits equal the costs. If a project's BCR is less than 1, the project's costs outweigh the benefits and it should not be considered. Therefore, our OFSP production which has BCR of 1.43 is a good investment.

We can still use Return on Investment (ROI) to know if the enterprise will be worth going into.

ROI = Net Return on Investment/Cost of Investment x 100%

= 506,000/352,400 x 100%

= 1.435 x 100%

= 143.5%

The above is one cycle production of 3 to 4 months and at the least plant sustainable potato production. Fresh healthy vines from a vine multiplication farm guarantees improved yield. The food producer farmer can effectively plan his planting season any time of the year.

We will look at the procedures for vine multiplication.

Preservation and Multiplication of Planting Materials.

As earlier stated, even though OFSP can be planted any time of the year as long as there is provision of water for the crop within the first 6 weeks, for most people, there is only one cropping season for it viz: plant in May/June and harvest about September. So, how does one preserve the vines from previous harvest for the next season?

Create a nursery plot nearby which you can effectively monitor and water through the dry season. Cut two nodes planting vines and plant in the nursery plot.

The Nursery Plot: The nursery plot should be a land that has not been planted potato. This is to ensure a potato-pest and disease free place. It must be a sand loamy soil, flat or with gradual slope. This is important for effective water management, to avoid water flowing away. It should be properly cleared, stumped and every root or debris removed to make for easy development of the vines. It is to be prepared in beds and not ridges. The land should be properly watered and made soft before planting. A plot of 100 square meters will be needed for producing vines for one hectare.

The Planting Vine Cutting: The selected planting vine should be disease-free, fresh and healthy. The planting vine is to be cut with just two nodes. To avoid wasting vines, it is therefore, recommended that if you are buying the vine it should be supplied in this cutting size and not the regular cut of 4-5 nodes; otherwise it should be delivered in its long strands and thereafter you cut in the size for your nursery. By this, the quantity of vine for the 100 square meter land will be quit minimal when compared to the quantity for production.

Planting of Nursery: Irrigate the beds and plat the 2-node vines on beds at the spacing of 20cm by 60cm. After 15 days of planting, apply Urea at the rate of 1.5kg/100 square metre. Irrigate the nursery as often as necessary to avoid dehydrating the young plants. After 6 weeks, the vines growing in the nursery can be cut off with sharp knife for planting in other nurseries (secondary nursery) or main plot. If on a secondary plot, this should be a larger land of about 500square metres and prepared in ridges at a distance of 60cm apart. The vine is still cut in two nodes and planted in 25cm spacing along the ridges at a depth of 2-3cm and covered with soil. On a secondary nursery, apply a urea (fertilizer) dressing of the plot at the end of the second week and the fourth week at the rate of 2.5kg each dressing. Like in the first (primary) nursery, at the end of the 6th week, the vines are ready for cutting for the main plot. This time the vine is cut in 3 - 4 nodes or 4 -5 nodes.

Weed Management of the nursery: The nursery will experience weed infestation as in main plot. Therefore, weeding is very important to ensure proper development of the nursery.

Calculating the cost benefit of vine multiplication: When each nursery vine is cut for planting the potato crop is still left in the ground to keep growing. Therefore, the first nursery of 100 square metres from where vine cuttings where taken to establish the second nursery of 500 square metres is ready by the 12th week for final harvest of both the tubers as well as the second set of vines. Still, vine cuttings for 1 hectare farm is available in the 500square metre nursery by that 12th week. The primary nursery can be allowed to stay up to week 15 before being harvested and prepared for another crop while the second nursery is left to the 5th month for final harvest of both the tubers and the second set of vines. The tubers from each of the nurseries may not be large in sizes but will be reasonable for sales. It must be mentioned that potato crop produces more branches once the first stem has been cut of. This is why it is easy to get the quantity of vines estimated here. The sales of vine from the both the first nursery and the second nursery within a period of 3 – 5 months gives great returns. Below are tables and calculations to show the profitability of vine multiplication enterprise.

Note: The farmer should acquire a land of not less than 600 square metres in size for the Vine Multiplication business in which he will make two nurseries.

Table 1
Nursery 1 Land size 100m^2

S/NO	DESCRIPTION	QUANTITY REQUIRED (Stands)	AMOUNT
1	VINE CUTTINGS (2Nodes length)	800	5,600
2	UREA FERTILIZER	1.5kg	200
	FERTILIZER APPLICAT ION	1 man day	1,500
3	WATER USE (3 months)	2000litres	800
4	LAND PREPARATION	1 man day	1,500
5	WEEDING	1 man day	1,500
5	Cutting and packaging vines 1 & 2	4 man day	4000
	TOTAL		**15,100**

Source: Simdozi International Ltd Farm Records 2020

Table 2

Nursery 2 Land size 500m^2 (Established 6 weeks after Nursery 1)

S/NO	DESCRIPTION	QUANTITY REQUIRED (Stands)	AMOUNT
1	VINE CUTTINGS (2Nodes length)	3,330	23,310
2	UREA FERTILIZER (1 & 2)	5kg	600
3	FERTILIZER APPLICATION (2 times)	2 man day	3,000
4	WATER USE (3 months)	4000litres	1,600
5	LAND PREPARATION	2 man day	3,000
6	WEEDING	2 man day	3,000
7	Cutting and packaging vines 1 & 2	4 man day	8,000
8	COST OF HARVESTING NUR 1 & 2	6 man days	24,000
	TOTAL		**66,510**

Source: Simdozi International Ltd Farm Records 2020

Total Cost for vine multiplication within 5 months = N15,100 + N66,510 = N81,610.

Table 3

OUTPUT AND SALES

S/NO	DESCRIPTION	QUANTITY REQUIRED	UNIT PRICE/BUNDLE	AMOUNT
1	Nursery 1 VINE CUTTINGS OUTPUTt (2Nodes length)	3,330 stands (33.3 bundles)	500	*(16,650 X 2) = 33,300
2	Nursery 2 Vine Cuttings Output (4 nodes length)	25,000 (500 bundles)	500	250,000**
TOTAL VINE REVENUE				**283,300**
	Nursery 1 Tubers	100kg	150	15,000
	Nursery 2 Tubers	800kg	150	120,000
TOTAL REVENUE FROM TUBERS				**135,000**
GRAND TOTAL REVENUE				**(283,300+135,000) 418,300**

Source: Simdozi International Ltd Farm Records 2020

*Nursery 1 vines are harvested first at week 6 to establish Nursery 2 and finally at week 12 when the tubers were also harvested.

**The vines from Nursery 2 during the final harvest at week 17/18 will be as much or even more than that of week 12 and will be sold just like the first.

true. You need much less resources to do vine multiplication and have more ready market for your output. So, this is a very good area for anyone who is in an area where OFSP is still at the introductory stage can get involved particularly where availability of land is a challenge.

References

1. Joshua Ombaka Owade, George Ooko Abong, and Michael Wandayi Okoth of University of Nairobi: **Production, Utilization and Nutritional Benefits of Orange Fleshed Sweet Potato (OFSP) Puree Bread- A Review**

2. *Dr Petros Nyakunu of PORENet Mozambique:* **Combatting Poverty With the Orange-fleshed Sweet Potato**

3. Joshua Ombaka Owade, George Ooko Abong, and Michael Wandayi Okoth of University of Nairobi: **Production, Utilization and Nutritional Benefits of Orange Fleshed Sweet Potato (OFSP) Puree Bread- A Review**

Orange-Fleshed Sweetpotato: Production, Processing and Utilization A Community Training Manual, Helen Keller International, Nigeria and International Potato Centre, Nigeria ,February, 2015

Benefit Cost Analysis of Orange Fleshed Sweet Potato (Ipomoea batatas L.) Varieties under Varying Planting Density by Donald A. UZOIGWE1, Comas O. MUONEKE1, Charles C. NWOKORO, Chikezie O. ENE, of Michael University of Agriculture Umudike, College of Crop and Soil Sciences, Department of Agronomy, P.M.B 7267, Umuahia,: www.notulaebiologicae.ro

Total Cost   = 81,610

Total Revenue   = 418,300

P/L   = 418,300 − 81,610 = 336,690

Total Net Benefit = N336,690

% Gross Margin Analysis = Net Benefit/Total Cost x 100%

$$= 336,690/81610 \times 100 = 4.12$$

$$= 412.55\%$$

Using Benefit Cost Ratio (BCR) to check the viability of the business:

The Benefit Cost Ratio (BCR) is given as: BCR = $\sum B / \sum C$ Where $\sum B$ = Total net benefit and $\sum C$ = Total cost of production of the enterprise.

BCR = Total net benefit/ Total cost of production of the enterprise.

= 336690/81610

= 4.12

Note: The methodology of interpreting benefit cost ratio indicates that BCR greater than 1, means that the Net Profit Value (NPV) of the project benefits outweigh the Net Profit (NPV) Value of the costs. Therefore, the project should be considered if the value is significantly greater than 1. If the BCR is equal to 1, the ratio indicates that the NPV of expected profits equal the costs. If a project's BCR is less than 1, the project's costs outweigh the benefits and it should not be considered. Therefore, our OFSP production which has BCR of 1.43 is a good investment.

We can still use Return on Investment (ROI) to know if the enterprise will be worth going into.

ROI = Net Return on Investment/Cost of Investment x 100%

= 336690/81610 x 100%

= 4.12 x 100%

= 412.5%

From the above demonstration and analysis, it looks unbelievable, but it is

1.   Boiled,

2.   Steamed

3.   Fried,

4.   Roasted,

5.   Dried

6.   Made into flour

7.   Made into a drink/juice

8.   Prepared alone, or combined with other food items.

S/NO DISH   **RECIPE**

1. Boiled OFSP with OFSP vegetable sauce

OFSP Tuber boiled with the skin, OFSP leaves, onion, fresh pepper, vegetable oil.

Note: The micronutrients especially Vit A in OFSP are best conserved and released into the body when boiled. This meal is good for all ages.

2. Boiled OFSP with fried egg/egg sauce

OFSP Tuber boiled with the skin, eggs, onion, fresh pepper, vegetable oil.

Note: This is a variant of the first but without OFSP vegetable sauce. It is more appealing to younger people.

OFSP Tuber, White Fleshed Sweet Potato, Irish Potato, choice vegetable.

3. Fried OFSP and other Potato chips with vegetables

61

Chapter 10

VARIOUS ORANGE-FLESHED SWEET POTATO DISHES

Orange-fleshed sweet potato contains some very critical micronutrients that endear it to all strata of the demography. Vitamin A is a very important micro-nutrient which we all need to be healthy. And this is obtainable in a good quantity in OFSP. That is why different persons and ages eat OFSP in different forms. There are so many ways of preparing OFSP for human consumption. Discussing this at this time is very important due to the rising severe acute malnutrition (SAM) in some parts of the world particularly Sub-Saharan Africa and South Asia. As OFSP is an easy to grow and easy to prepare composite food crop which comes handy in fighting SAM, it is necessary that the reader pays attention to any OFSP dish that appeals to him/her in this guide.

Pregnant women and lactating mothers need a lot of Vitamin A for themselves and for their unborn or breastfeeding babies. Severe deficiency of this could be fatal to the baby leading to malformation of the fetus, miscarriages, premature arrival of baby; and night blindness for the woman. And children 0 - 5 years of age lacking Vitamin A have immunity problem and so run higher risk of infections, poor cognitive development, stunted growth and death.

At a time the world is challenged by a pandemic virus disease, building up one's immunity cannot be over emphasized. 100 grams of OFSP supplies you your daily Vitamin A body need.

In this Chapter therefore, we are going to look at some of the various dishes made with OFSP. There are quite a number of dishes and every day as many more people and cultures adopt OFSP they also adopt it to their cultural meals. OFSP can be prepared in these different ways:

4. Boiled OFSP and other Potatoes with vegetables

OFSP Tuber, White Fleshed Sweet Potato, Irish Potato, choice vegetable.

This a variant of the above except that the potatoes are boiled instead of fried and eaten with vegetable sauce. This is a composite food as it combines all the peculiar nutrients of the different varieties of potatoes.

OFSP Tuber, salt, vegetable oil.

This is a very easy to prepare snack.

5. Fried OFSP Chips

OFSP Porridge

OFSP Tuber, salt, vegetable oil.

This is a very easy to prepare meal.   Best when spiced with scent leaves vegetable or any other spicy vegetable. Very filling but does not make one feel heavy.

Rice and Beans with OFSP

OFSP chopped in small sizes, Rice and Beans boiled together.

The OFSP cubes are added 5-7 minutes before the meal is ready. This is an appealing way of eating your OFSP without the consciousness of eating potato along with the rice and beans

8. OFSP Soup with Egusi (Melon)

OFSP Vegetable, Melon, other ingredients for making soup.
This is popular in Nigeria and other places where eating Garri or other meals with soup is major part of their culture.

OFSP Juice/Smoothie

OFSP tuber alone or with other fruits/vegetables.
This is done with either raw tuber or boiled tuber.

10. OFSP Flour

OFSP Tubers, sliced, sun-dried and ground into flour

Mashed OFSP Meal

OFSP Tuber boiled and mashed. Best for weaning babies.

Summary

1. Orange Fleshed Sweet Potato is easy to grow. It can grow on any soil – marginal – simple sandy loamy soil. It does not need a lot of care or even fertilizer to grow and produce. It can be grown in pots or non-land medium.

2. The planting material is cheap. You can use the vines or the tubers cut into smaller portions.

3. It has a short cycle – ready for harvest within 3 months – 90 days.

4. Every part of the OFSP is useful. The tuber can be used to prepare all manner of dishes; the leaves used for soup, sauce, salad – and the vine from where the leaves were plucked planted back to start the cycle again; the leaves, and the tuber peels are good meals for livestock-rabbits, goats, pigs, cattle, and even chicken.

5. It is easy answer in fighting malnutrition – highly nutritious and medicinal. OFSP is being promoted as a nutrition intervention to tackle Vitamin A Deficiency (VAD) and food insecurity in many countries.

6. It is a delicious and healthy snack for the children and the entire family.

7. The popularity of OFSP is still growing. So, it has great business and wealth creation potentials. Tubers can be grown and supplied to regular markets and chain stores; or the vines for farmers (as the vine is not easily available). OFSP can also be produced for processing and value addition, or just concentrate on marketing OFSP products.

8. Return on investment (ROI) is very high. For tuber production it is 143% per cycle of 3 months. For Vine production, the ROI is 412.5%!

9. OFSP has a lot of potential for further research and development.

10. It can be prepared into very many delicious dishes.

There is no doubt that OFSP is one of the most nutritious crops with a lot of potential yet to be researched and discovered.

About the authors

Hon Dr **John Abraham Godson** is a graduate of Agriculture from the Abia State University Uturu (1987-1992). He also holds 4 Masters and 2 doctorate degrees. He was a Research supervisor at the International Institute of Tropical Agriculture- IITA Ibadan. He has lectured in various universities in Europe and is the first black member of the Polish Parliament in Polish history (2010-2015). He is currently the CEO of Pilgrim Ranch Nigeria, an integrated allied farm functioning in various aspects of the agricultural value chain. He is also an investor, social entrepreneur, author, consultant, mentor and coach.

Esonu Nduka Udeala Has Bachelors' degree from University of Nigeria Nsukka 1989; and PGD and MBA in Business Administration and Marketing from University of Nigeria Enugu Campus. He was further trained in Agribusiness and Value Chain Development. He joined National Fadama Development Project, a World Bank assisted agricultural project implemented in the 36 States and Federal Capital Territory of Nigeria from 2010 to 2019 where he served in various capacities as Community Development Officer, Agribusiness Investment and Management Officer. He is a Consultant to several international donor agencies and agricultural organizations including IFAD/Value Chain Development Program (VCDP), CSS Group Gora Nasarawa State, Nigeria Incentive-Based Risk Sharing System for Agricultural Lending (NIRSAL). He is presently the Lead Consultant for Simdozi International Ltd a nutritional agricultural company that has been promoting the production and utilization of food crops rich in Vitamin A, C, B, zinc, iron and other essential micronutrients in order to fight the prevailing severe acute malnutrition (SAM) in Nigeria. He championed the establishment of Orange-Fleshed Sweet Potato (OFSP) farms in some states of Nigeria including Adamawa, Imo, Cross River, Lagos and Ogun; and is still leading the campaign on it becoming a major staple in the country.